For Margot Clementine, Isaiah, Amaya, and
all the others yet to fledge —CC

To my mom, for always championing my
creativity and encouraging me to let my
dreams be my wings —KA

About This Book

The illustrations for this book were done digitally using Procreate. This book was edited by Erika Turner and designed by Tracy Shaw. The production was supervised by Patricia Alvarado, and the production editor was Marisa Finkelstein. The text was set in Alegreya, and the display type is Bubblegum Sans.

The Urban Owls

How Flaco and Friends Made the City Their Home

By
CHRISTIAN COOPER

Illustrated by
KRISTEN ADAM

Once there was an owl named Flaco, who flew the sky at night. Some owls are big, some of them small. But Flaco? He was the biggest of all!

Owls in the city can come from near or far, like the tundras of the north or the woods in the south. But the ones we know best all landed in the same place: New York, where buildings reach so high they touch the sky!

It's a very big city full of people and their dreams.

In a city so full, is there room for owls? One night, Flaco decided to find out.

He had spent most of his life in the Central Park Zoo, where there wasn't much room to fly and there wasn't much to do.

But then Flaco got loose, and he flew and flew and flew!
If you'd been stuck in a cage your whole life, wouldn't you?

The rush of the city surrounded Flaco.

Some people worried that he might not survive. But Flaco surprised everyone—he could do just fine on his own. He used his great big owl eyes to see in light much too dim for you and me.

So Flaco worked the night shift, like most owls do (and like some people do too). In the dark he caught prey, and he slept during the day.

And even though his family was from very far away—a place
called Eurasia, near Africa—Flaco became a New Yorker.
And that's the way it is with a lot of people too.

Just like Flaco, Barry the Barred Owl found her place in the city. And like so many other wild creatures, she made Central Park her home.

Barry enjoyed it so much, she stayed longer than any Barred Owl before! She found out there was great food in town, and she knew how to get it.

Her wings had special owl feathers that made them *silent*. Nobody could hear Barry's wings flap! Her prey never heard her coming.

Near a winding path, she had a favorite roost, which is a place where a bird likes to perch and rest. Barry roosted there so often that curious people knew where to go to see her.

Sometimes she would stare back with her big dark eyes.

From her perch at her roost by that path in the park,
can you imagine all the things she could see?

Geraldine the Great Horned Owl lives in Central Park too. It seems for an owl in the city, living in the park is the thing to do!

Her kind of owl looks like Flaco but has always lived in America. Big and strong and fierce, these owls rule the woods.

Female owls are bigger than male owls, so Geraldine is even bigger, stronger, and fiercer than the Great Horned Owl guys!

But one day she hurt her foot, and it never healed right.

Geraldine doesn't let that stop her! She still catches all the food she needs.

A body that's different doesn't mean you're not able to do great things.

Just ask Geraldine!

Sometimes visitors come to the city just for a little while. One tourist from the cold north caused a lot of excitement: *a Snowy Owl*. When she showed up in Central Park, she was the first Snowy Owl there in more than a hundred years!

Because this owl is big and white and hunts during the day, she was easy to see. People lined up with cameras to snap her picture.

In New York, being unusual can be *fabulous*! But her visit was so short, nobody had a chance to give her a name.

What would you name her?

The big city isn't for everyone. Rocky the Saw-whet Owl lived many miles away in the quiet woods. One day, she went to sleep in her favorite pine tree. It was seventy-five feet tall and always had branches to hide in. That way no one would bother her while she slept.

But when she woke up, the tree had moved! Someone
had chopped it down and driven it to the city. It ended up
in Rockefeller Center, where it would soon be decorated
with lights for everyone to see.

No one knew Rocky was still there!

Sirens!

Shouts!

Chaos!

Crowds!

The harsh sounds of the city
surrounded Rocky.

She turned her owl neck nearly all the
way around to look one way . . .

. . . then turned it nearly all the way
around to look the other way.
But Rocky didn't know where she was.
This wasn't her quiet woods!

But some kind people helped
Rocky get far away from the city
and back to the woods.

Most people think owls live only in the countryside, like Rocky does. But Flaco, Barry, Geraldine, and many other owls have called New York City—and other cities—their home.

They live in our parks, our gardens, our own backyards.

So wherever you live, even in a city full of people and their dreams, look for our neighbors the owls. Because they have a place here too.

More Facts About Flaco and Other Famous Owls

Eurasian Eagle Owls like Flaco are normally found in Europe and Asia. They prefer rocky habitats but are highly adaptable. Their call is characterized by low hoots that are spaced out. You can sing along too! Try: *hooo…hooo…hooo…hooo!*

Barred Owls like Barry are found in North America and live in densely wooded habitats. Their name comes from the bar-like stripes on their underside. The feathers under Barred Owls' wings turn pink when they eat a lot of crayfish! Their call is characterized by loud, barking hoots: *hoo, hoo, hoo-hoo; hoo, hoo; hoo, hoo-aw!*

Great Horned Owls like Geraldine are one of the most common owls in North America and can be found in all types of habitats. They have plumicorns (tufts of feathers that sit on top of their heads) that look like horns. A Great Horned Owl's call is characterized by a series of low, resonant hoots: *hoo, hoo-hoo, hoo, hoo!*

Snowy Owls live in the most northern parts of the world. They are white with some dark markings. While females keep some of their markings for life, male Snowy Owls get whiter as they get older. Snowy Owls are usually silent, but when they do sound off, their call is closer to loud, slightly rasping hoots: *hoo-hoo!*

Northern Saw-whet Owls like Rocky are found in forests across North America. About the size of a Robin, they are some of the smallest owl species in North America. Like Snowy Owls, Northern Saw-whet Owls are usually silent. When they do make sounds, their call is a series of high-pitched toots: *Toot! Toot! Toot!*

DID YOU KNOW?

- Owl eyes are adapted to see with very little light. Some owls can hunt in a football field lit by a single candle.
- Owl wings have special feathers on the leading edge that make them silent when flapping.
- Owl necks have fourteen bones—twice as many as human necks have!—which lets owls turn their heads 270 degrees (almost a full circle!) in either direction.

HOW TO SPOT AN OWL

Looking for birds, especially owls, is a fun and exciting way to interact with nature. Here are some tips to get started.

- **RESEARCH:** Check out birding and nature books to learn more about birds and where to observe them. With an adult's support, you can also find out if there are youth birding groups in your area where you can exchange tips, bird pics, and sighting locations with other young birders.

- **LISTEN:** You'll be far more likely to hear an owl before you see one! Find trusted websites or apps that provide audio where you can familiarize yourself with the sounds. During the day, listen for jays and other birds making a fuss; often that's because they've spotted an owl, and they're alerting their neighbors that there's danger!

- **LOOK:** Together with an adult, scope out wooded areas close to clearings or fields. Many owl species hunt in open areas and bring their prey back to their perch or nest. If you spot a mess of feathers on the ground, you may have found a place where a smaller bird was preyed upon. Look out for owl pellets (bones and fur that the owl can't digest, so it spits them back out as a ball)—they are a sign that owls could be close by.

HOW TO BE A GOOD OWL NEIGHBOR

Owls in cities face many dangers, but one of the worst is the use of rodenticide (a poison) to kill rats and mice. If an owl eats a poisoned rodent, the owl gets poisoned too, and it may get sick, become weak, or even die. Talk to people in your community about using other, safer measures to control rat and mice populations. The owls will thank you!

FURTHER READING

Websites

The Peregrine Fund: https://peregrinefund.org/

National Geographic Bird Pictures & Facts: https://www.nationalgeographic.com/animals/birds

Books

Burgess, Murry, and Tamisha Anthony (illustrator). *Sparrow Loves Birds*. New York, NY:
Little, Brown and Company, 2024.

Editors of Storey Publishing. *Backpack Explorer: Bird Watch: What Will You Find?* New York, NY:
Storey Publishing, 2020.

Belleny, Danielle, and Michelle Carlos (illustrator). *Junior Birder's Handbook: A Kid's Guide to Birdwatching*.
New York, NY: Running Press, 2023.

SELECTED SOURCES

"Barred Owl." Audubon. Accessed May 2, 2024. https://www.audubon.org/field-guide/bird/barred-owl.

Dhanesha, Neel. "10 Fun Facts about the Barred Owl." Audubon. February 12, 2021.
https://www.audubon.org/news/10-fun-facts-about-barred-owl.

"Eurasian Eagle-Owl." The Peregrine Fund. Accessed May 6, 2024. https://peregrinefund.org
/explore-raptors-species/owls/eurasian-eagle-owl.

"Great Horned Owl." All About Birds, Cornell Lab of Ornithology. Accessed May 6, 2024.
https://www.allaboutbirds.org/guide/Great_Horned_Owl/overview.

"Great Horned Owl." Audubon. Accessed May 2, 2024. https://www.audubon.org
/field-guide/bird/great-horned-owl

"Northern Saw-whet Owl." Audubon. Accessed May 6, 2024. https://www.audubon.org
/field-guide/bird/northern-saw-whet-owl.

"Snowy Owls & Airports." NYC Bird Alliance. Accessed May 6, 2024. https://www.nycbirdalliance.org
/our-work/conservation/urban-raptors/snowy-owls-and-airports.